Growing Up

Tadpole to Frog

by Brenna Maloney

EXPLORE the LIFE CYCLE!

Content Consultant

Lisa Monachelli, Education Director

Stamford Museum & Nature Center

SCHOLASTIC

Library of Congress Cataloging-in-Publication Data
Names: Maloney, Brenna, author.
Title: Tadpole to frog / Brenna Maloney.
Description: New York: Children's Press, an imprint of Scholastic Inc.,
   2021. | Series: Growing up | Includes index. | Audience: Ages 6-7. |
   Audience: Grades K-1. | Summary: "Book introduces the reader to the life
   cycle of a frog"— Provided by publisher.
Identifiers: LCCN 2020031791 | ISBN 9780531136980 (library binding) | ISBN 9780531137093 (paperback)
Subjects: LCSH: Frogs—Life cycles—Juvenile literature.
Classification: LCC QL668.E2 M2946 2021 | DDC 597.8/9—dc23
LC record available at https://lccn.loc.gov/2020031791

Produced by Spooky Cheetah Press. Book Design by Kimberly Shake.
Original series design by Maria Bergós, Book&Look.

Printed in Heshan, China 62

SCHOLASTIC, CHILDREN'S PRESS, GROWING UP™, and associated logos are
trademarks and/or registered trademarks of Scholastic Inc.

1 2 3 4 5 6 7 8 9 10 R 30 29 28 27 26 25 24 23 22 21

Scholastic Inc., 557 Broadway, New York, NY 10012.

Photos ©: Photos ©: 7: Michel Loup/Biosphoto/Minden Pictures; 9: Westend61/Getty Images; 10 top: Wild
Horizons/Universal Images Group/Getty Images; 10 bottom: Gerard Lacz/FLPA/Minden Pictures; 10 background:
Paul Broadbent/Alamy Images; 12-13, 14-15: Kubo Hidekazu/Nature Production/Minden Pictures; 18: D. Parer and
E. Parer-Cook/Minden Pictures; 21: Kim Taylor/NPL/Minden Pictures; 22 bottom left: André Simon/Biosphoto; 22
branch: Freepik; 23 foreground: Carl Monopoli/Dreamstime; 24: David & Micha Sheldon/Getty Images; 26 center:
Arto Hakola/Getty Images; 26 bottom: MYN/Javier Aznar/NPL/Minden Pictures; 27 top: Ernst Dirksen/Buiten-
beeld/Minden Pictures; 27 center left: Michael and Patricia Fogden/Minden Pictures; 27 center right: Cynoclub/
Dreamstime; 27 bottom: Cyril Ruoso/NPL/Minden Pictures; 29: Jelger Herder/Buiten-beeld/Minden Pictures; 32
center: Nature Picture Library/Alamy Images.

All other photos © Shutterstock.

Table of Contents

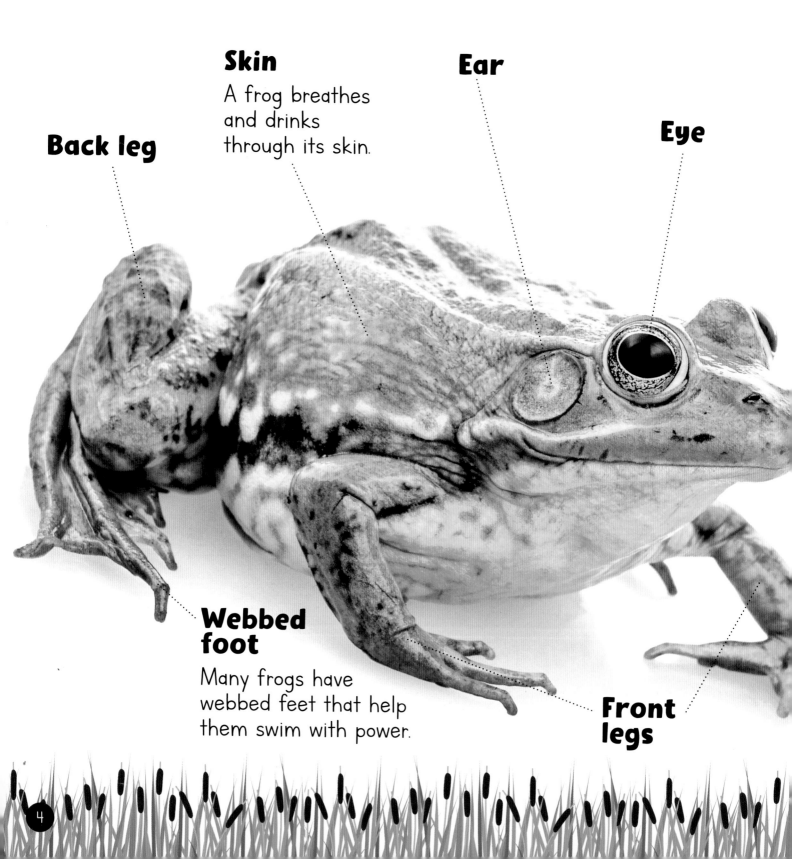

Back leg

Skin
A frog breathes and drinks through its skin.

Ear

Eye

Webbed foot
Many frogs have webbed feet that help them swim with power.

Front legs

Fantastic Frogs

A frog is an amphibian (am-FIB-ee-uhn). *Amphibian* means "two lives." Frogs begin their lives in the water. When they are adults, they live on land. Frogs change as they grow. That process is called **metamorphosis.**

There are many types of frogs. Some are smaller than a coin. Others are as big as a cat!

It Starts with an Egg

The time it takes to grow from a tadpole to a frog is different for every **species**. But all frogs begin their lives the same way: inside a tiny egg. Female frogs of most species lay their eggs in water. Inside each egg are **cells** that will grow into a tadpole.

A female frog can lay thousands of eggs at a time.

The black dot in each of these eggs is the beginning of a tadpole.
▼

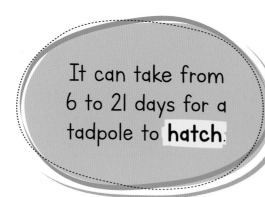

It can take from 6 to 21 days for a tadpole to **hatch**.

Tales of a Tadpole

A tadpole looks a bit like a fish. It doesn't have legs. It has just a mouth, gills, and a tail. Gills help the tadpole breathe underwater. At first, the tadpole sticks close to floating weeds or grasses. Later, it starts to swim around and feed on **algae**.

Sometimes young tadpoles swim together in schools, the way fish do.

Older tadpoles
sometimes eat younger tadpoles.

Fish
like this rudd can gobble up many tadpoles.

Insects
like this great diving beetle larva eat tadpoles that they catch in the water.

Tadpoles, Beware!

The early stages of a frog's life can be very dangerous. For tadpoles, the risk of being eaten is very high. This is especially true right after they hatch. Hiding among water plants helps them stay safe from **predators** like fish, birds, and even insects.

Time to Change

A tadpole develops lungs, which will enable it to breathe air. Depending on the species, that can take days or weeks. As the tadpole grows teeth, it can eat small insects instead of just plants. The tadpole also grows back legs, so it can kick when it swims.

The back legs form first on a tadpole.

Like an adult frog's, a froglet's two front legs have four toes each. The back legs have five toes each.

▼

Like an adult frog, a froglet has eyes on top of its head. The frog can see in almost every direction.

Becoming a Froglet

Next, the tadpole becomes a froglet. Its body becomes more like that of an adult frog. The froglet's lungs get bigger, and its gills disappear. The shape of its head changes, and its eyes start to bulge. The froglet grows thin front legs with skinny fingers. Its tail starts to shrink.

Into Adulthood

When its tail finally disappears, the frog is ready to leave the water. It hops on land and breathes air. Adult frogs spend a lot of time on land. But most frogs need to stay near water or wet places. If they don't, their skin will dry out.

A thin layer of skin covers a frog's earholes. It keeps dirt and water out.

This water-holding frog is getting rid of its old skin.

Skin-Deep

Frogs don't drink water. Any moisture they need passes through their skin. Their skin can also absorb oxygen from water. That enables them to breathe underwater.

A frog sheds its skin often. It wriggles out of the old skin, pulling it over its head like a sweater! Often, the frog eats the skin.

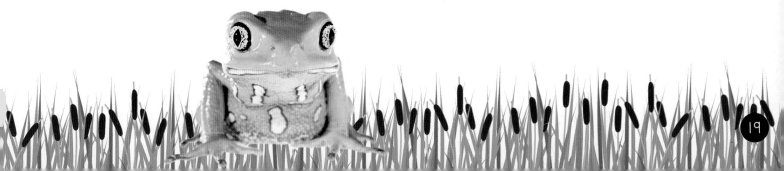

Sticky Situation

Frogs are **carnivores**. Small frogs eat insects such as flies, moths, and dragonflies. Larger frogs eat grasshoppers and worms. Almost all frogs use their sticky tongues to catch a meal. The tongue snaps out and wraps around the **prey**. Then it throws the prey into the frog's mouth. The frog blinks as it swallows. That presses the frog's eyeballs onto its mouth to push the food down its throat.

A frog can snatch an insect faster than you can blink!

A frog's tongue is attached to the front of its mouth.

Frog Predators

Adult frogs have predators, too. They must hide from birds and reptiles. Their bodies are often dark on top and light on the bottom. When seen from above, the frog's back blends in with the dark water. When seen from below, its light belly blends in with the sky.

Snakes, lizards, and other **reptiles** eat frogs on land.

Birds like this heron snatch frogs and froglets from the water.

A male frog inflates his vocal sacs to ▶ attract a mate.

Some frog calls are so loud they can be heard a mile away.

Finding a Partner

When frogs are two or three years old, they look for a **mate**. Male frogs call out to female frogs. Each species of frog has its own special call. They croak, squeak, chirp, and grunt. A female chooses a male. Then she finds a spot where she will lay her eggs. The frog pair mates—and the life cycle begins again!

Frog Facts

A group of frogs is called an army.

Frogs are ectothermic. That means their body temperature changes with the temperature of their surroundings. Frogs live on every continent except Antarctica.

The glass frog is see-through! You can look through its skin to see its heart beating and its stomach digesting food.

Many frogs can leap more than 10 times their body length.

The waxy monkey frog oozes a wax that it rubs over its body. The wax acts like sunscreen.

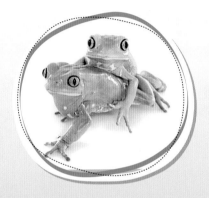

When Darwin's frog tadpoles hatch, the male frog swallows them. He keeps them in his vocal sac for about 60 days while they grow. Then he opens his mouth, and fully formed frogs hop out!

The African clawed frog doesn't have teeth or a tongue. It uses its hands to push food into its mouth.

Growing Up from Tadpole to Frog

A frog goes through complete metamorphosis as it grows. The time it takes to grow from a tadpole to a frog is different for every species.

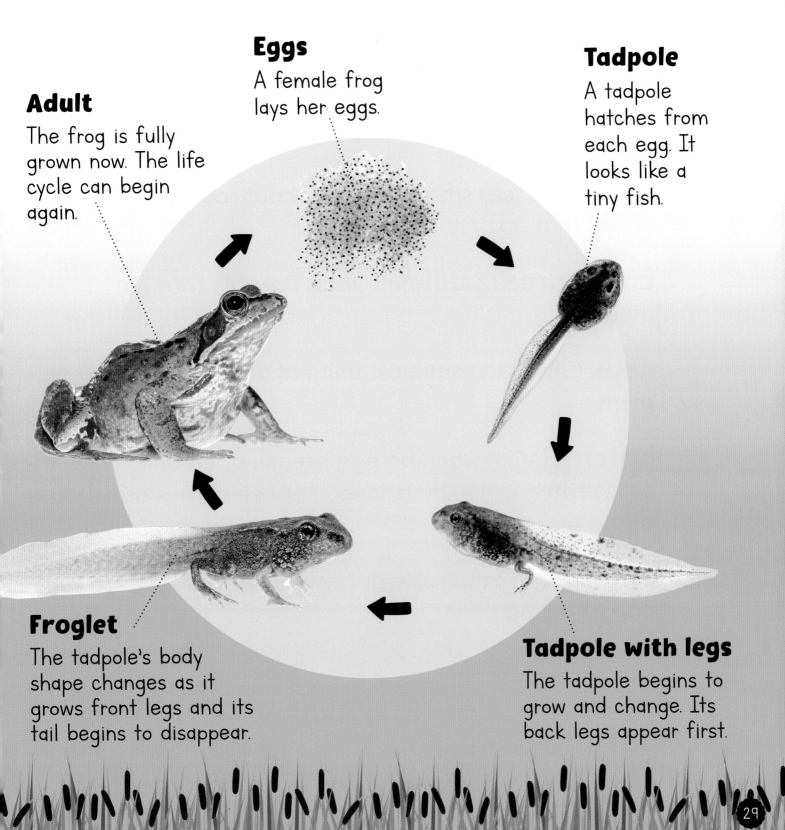

Adult
The frog is fully grown now. The life cycle can begin again.

Eggs
A female frog lays her eggs.

Tadpole
A tadpole hatches from each egg. It looks like a tiny fish.

Froglet
The tadpole's body shape changes as it grows front legs and its tail begins to disappear.

Tadpole with legs
The tadpole begins to grow and change. Its back legs appear first.

Glossary

algae (AL-jee) small plants without roots or stems that grow mainly in water

carnivores (KAHR-nuh-vorz) animals that eat meat

cells (SELZ) the smallest units of an animal or a plant

hatch (HACH) when an egg breaks open and a bird, reptile, amphibian, insect, or fish comes out of it

mate (MATE) the male or female partner of a pair of animals

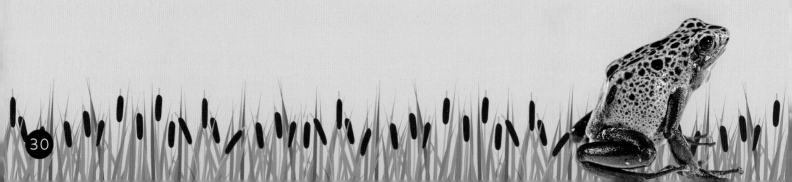

metamorphosis (met-uh-MOR-fuh-sis) a series of changes some animals go through as they develop into adults

predators (PRED-uh-turz) animals that live by hunting other animals for food

prey (PRAY) an animal that is hunted by another animal for food

species (SPEE-sheez) one of the groups into which animals and plants are divided

Index

About the Author

Brenna Maloney is a writer and an editor. If she could be a frog, she would want to be a desert rain frog because they look ridiculous and they sound like a dog's squeak toy. Look them up! Maloney lives and works in Washington, D.C., with her family.